LED LIGHTING

A review of the current market and future developments

Hilary Graves and Cosmin Ticleanu

bre press

ACKNOWLEDGEMENTS
The authors of this report would like to thank the following BRE colleagues for their contributions:
• Matt Blaikie
• Paul Littlefair
• Chloe Murphy
• Andrew Thorne
• Sam Woods

PUBLISHER'S NOTE
All URLs accessed 9 September 2011. The publisher accepts no responsibility for the persistence or accuracy of URLs referred to in this publication, and does not guarantee that any content on such websites is, or will remain, accurate or appropriate.

Cover images:
Main: LED lighting with good colour rendering for medical purposes
Top right: LEDs for automotive applications (courtesy of Osram)
Middle right: LED lamp with heat sink
Bottom right: Adjustable LED for accent lighting (courtesy of Osram)

CONTENTS

EXECUTIVE SUMMARY

Lighting is an essential part of everyday life in the developed world and is one of the largest single users of energy, being responsible for between 15% and 22% of all electricity use in buildings. Good lighting is considered essential to health, well-being and productivity, but the efficiency of common light sources can vary widely.

With demands from the UK government and international agreements to reduce carbon emissions, building designers, owners and occupiers are looking at the energy efficiency of their lighting. Not since the late 1970s has there been such a focus on energy management in lighting.

Light-emitting diodes (LEDs) are a proven technology that offers enormous possibilities for providing highly energy-efficient and good-quality lighting. This BRE Trust Report summarises the current LED market and various advantages of LEDs, and outlines the challenges and barriers to widespread adoption of the technology.

(Courtesy of Switch Lighting)

1 INTRODUCTION

Lighting is an essential part of everyday life in the developed world and is one of the largest single users of energy, being responsible for between 15% and 22% of all electricity use in buildings. Good lighting is considered essential to health, well-being and productivity, but the efficiency of common light sources can vary widely.

With demands from the UK government and international agreements to reduce carbon emissions, building designers, owners and occupiers are looking at the energy efficiency of their lighting. Not since the late 1970s has there been such a focus on energy management in lighting.

Light-emitting diodes (LEDs) are a proven technology that offers enormous possibilities for providing highly energy-efficient and good-quality lighting. The technology was initially developed in 1907 but the first white LED was not produced until 1996. Since then, considerable effort has been, and continues to be, made to improve the efficacy of both LED and organic LED (OLED) lighting and to reduce the costs of manufacture so that LED light engines can be developed as commercially viable alternatives to conventional light sources.

If LED lighting achieves its expected levels of efficiency, then with high levels of uptake the energy consumption of domestic and commercial lighting could potentially be reduced by up to 70% by 2050. It could realistically achieve a 37% saving in lighting energy use by 2030[1].

A great deal of fundamental research and development of new components is ongoing, including development of new LED materials, especially in the green/yellow part of the spectrum, and development of all forms of OLED. The feasibility of the technology has been demonstrated, but LED products still need to be developed further before they can give energy savings comparable to those of competing types of lighting, and fully meet customer requirements for light output, colour and reliability.

2 BACKGROUND

2.1 LED LIGHT ENGINES

An LED light engine is a combination of one or more LED devices or arrays, a driver, heat sink and electrical and mechanical connections. It is intended to be mounted in a luminaire (ie a light fixture or fitting). The LED itself is a semiconductor material protected by an encapsulant that allows light to emerge. A schematic of a typical LED light engine is shown in Figure 1.

LEDs require a driver to convert mains power into the current and voltage required by the semiconductor. The driver may also sense and correct for changes in intensity and colour during operation.

A self-ballasted LED is a unit that cannot be dismantled without being permanently damaged. It is provided with a standardised lamp cap and incorporates an LED light source and any additional elements necessary for the start-up and stable operation of the light source.

Because the output and lifetime of LEDs are adversely affected by heat, a heat sink is usually provided. High-power LEDs that emit a large amount of heat are usually also placed in a special luminaire. Figure 2 shows LED lamps with different types of heat sink.

2.2 LED MANUFACTURE

LEDs are manufactured by a process known as 'epitaxy', in which crystalline films of semiconductor material are deposited onto a substrate. OLED devices made with small organic molecules are usually produced by vacuum deposition onto a substrate, generally glass. Polymer OLEDs (P-OLEDs) can be deposited by inkjet printing onto glass or a mechanically flexible substrate such as a polymer film. Addressable OLEDs, which allow individual pixels of material to be switched on separately, can also be fabricated in the laboratory in this way. These can be used for signage or screens[2].

General lighting devices can be based on a single LED, or alternatively several LEDs can be packaged together on a common substrate or wiring board to form an LED array. This may be done to increase total light output or modify the spectrum (eg using red, green and blue LEDs together to produce white light).

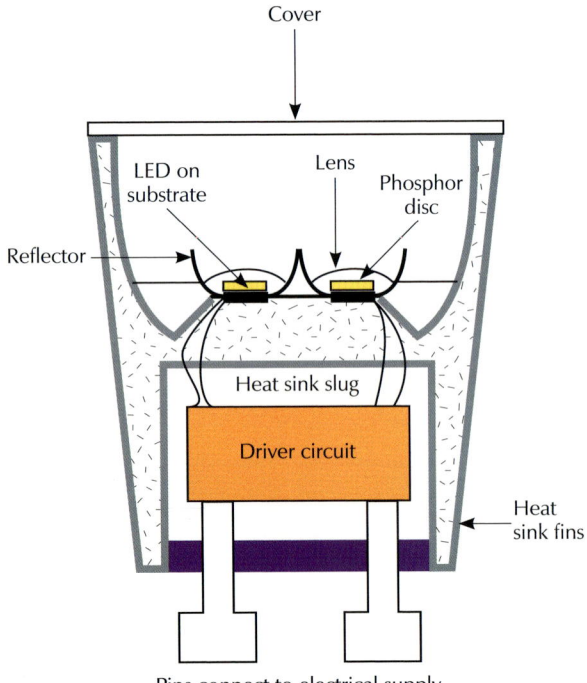

Figure 1: Schematic of a typical light engine

Figure 2: Two LED lamps showing different heat sinks

2.3 LED LIGHTING APPLICATIONS

LEDs are currently mainly used in niche applications, such as coloured decorative lighting; addressable picture walls; emergency lighting, automotive and aviation lighting; low-power display lighting; and LCD backlighting. Since LEDs are point sources they are capable of providing the kind of 'sparkle' often desired by lighting designers for leisure and retail commercial applications. Table 1 summarises the typical applications of the various types of lighting, while Figure 3 shows a breakdown of end uses for LED lighting.

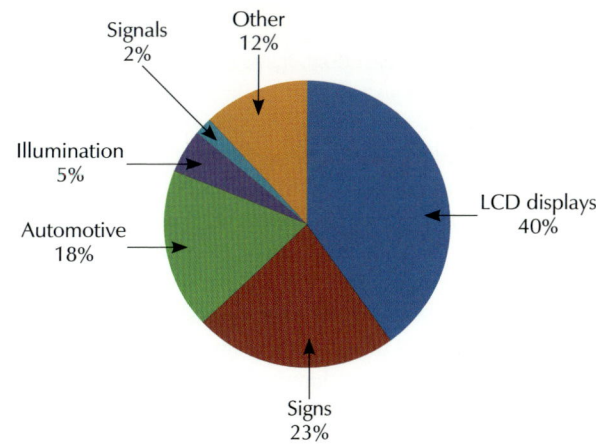

Figure 3: Market breakdown of LED use

Light source	Applications
Table 1: Light sources and typical current applications	
Tungsten filament	General domestic lighting Small-scale commercial Residential (hotels, residential care homes) Leisure (restaurants, pubs, clubs) Specialist Automotive and aviation Traffic signals
Tungsten halogen	General domestic lighting Small-scale commercial (downlighting and spotlighting) Display lighting in all types of building (especially retail and leisure sectors) Specialist Automotive
Tubular fluorescent	General commercial lighting Public sector (including schools and hospitals) Industrial and storage Communications and transport Domestic (kitchens and garages)
Compact fluorescent	General domestic lighting Commercial (smaller spaces, some general lighting) Public sector (including schools and hospitals) Communications and transport
Low-pressure sodium	Street lighting
High-pressure sodium	Street lighting Other external lighting including floodlighting Industrial and storage Some retail lighting
Metal halide	Retail lighting (displays or general lighting) General lighting in industrial and storage buildings Floodlighting
Induction lamps	Used where access is difficult, in some external lighting (tunnels, precincts) and large internal spaces like atria
Sulfur lamps	Large-scale industrial
LEDs	Coloured decorative lighting Addressable picture walls Emergency lighting Automotive and aviation lighting Low-power display lighting LCD display backlighting
OLEDs	Mobile phone displays Other small-display applications

Recent applications of LEDs include vehicle headlights (Figure 4), traffic signals, downlights (primarily for commercial accent and display applications), emergency lighting and decorative outdoor lighting, particularly in conjunction with small-scale solar panels.

Compared with other forms of lighting, LEDs tend to have lower light output and lower wattage: typically a few watts with currents in the order of milliamps or tens of milliamps. Since the light output of individual LEDs is small compared with that of conventional lamps, multiple LEDs are often used together to create high-power LEDs, which, due to their higher lumen output, can be used to replace other lamps. High-power LEDs can be driven at currents from hundreds of milliamps to more than an ampere and have significantly higher lumen output.

The best-performing commercially available warm white LED fittings have an efficacy of 50–60 lumens per watt (lm/W). Sometimes low-wattage LEDs are marketed as suitable replacements for halogen lamps. However, at the above efficacies a 12–14 W LED would be needed to provide the equivalent output of a basic 50 W halogen lamp, so lamps rated at a few watts provide correspondingly lower levels of light output. Some examples of the different types of LED alternative to halogen lamps are shown in Figure 5.

Within a few years, it is expected that the efficacies of LED lamps will rise to 100 lm/W or above (the highest-efficiency high-power white LED already achieves 115 lm/W), so lower-wattage lamps may then be able to provide the required amounts of light.

The spectrum of a light source will affect its colour appearance and colour rendering. LEDs and OLEDs are made from a variety of semiconductor materials in order that a range of colours or white light can be emitted (Table 2). LEDs are available but their colour qualities may differ from those of other lamp types (see Table 3 and Section 4 of this publication).

White LEDs with good colour rendering can be made by combining red, green and blue LEDs, but they are less energy-efficient. More efficient white LEDs can be made by coating a blue LED with a phosphor, but this has reduced colour-rendering capabilities. Work is continuing on improving the efficacy of multiphosphor coatings, which have better colour-rendering qualities. Where research is seeking to develop more efficient white light, the focus is on improving the energy efficiency and colour rendering of gallium nitride-based white LEDs.

LEDs normally last much longer than other forms of lighting – 50,000 hours compared with 10,000 hours for compact fluorescents. This and other comparisons are set out in Table 3. Further explanation of terms such as 'correlated colour temperature' is given in Section 4 and in the glossary of terms and abbreviations.

Figure 4: LED headlights
(Courtesy of Audi UK)

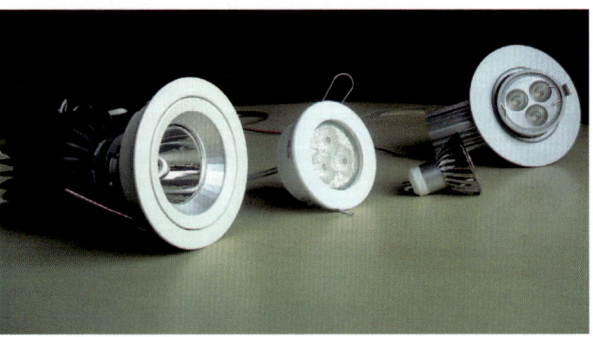

Figure 5: Different types of LED alternative to halogen lamps
From left to right: LED fixed downlight (31–45 lm/W); 350 mA self-contained cool white LED downlight (54–70 lm/W); GU10 white LED bulb (60 lm/W); and adjustable LED downlight (47 lm/W)

Table 2: Examples of semiconductor materials used in the manufacture of LEDs

LED colour	LED material
Infrared	Gallium arsenide
Red	Gallium arsenide phosphide
Orange	Aluminium gallium indium phosphide
Yellow	Gallium arsenide phosphide
Green	Gallium phosphide, aluminium gallium phosphide
Blue	Gallium nitride; silicon carbide and sapphire as a substrate
Ultraviolet	Aluminium gallium indium nitride
Incandescent white	Gallium nitride
Cool white	Gallium nitride

Table 3: Comparison of 2007 commercial LEDs and conventional light sources*[3]

Light source	Typical luminous output (lm unless indicated otherwise)	Typical wattage (W)	Typical luminous efficacy (lm/W)	Correlated colour temperature/ dominant wavelength (K unless indicated otherwise)	Colour rendering index	Typical lifetime (hours)†
LEDs						
Red LED	42	0.3	58	625 nm	N/A	50,000
Amber LED	42	0.8–1	50	590 nm	N/A	50,000
Green LED	53	1	53	530 nm	N/A	50,000
Blue LED	21	1.2	18	470 nm	N/A	50,000
White LED	220	4	55	4000	70	50,000
Warm white LED	180	4	45	3000	90	50,000
SM-OLED	1000 cd/m²		46[4]	~3000	80	5000
P-OLED	1000 cd/m²		6.5 (white)			5900‡
Conventional						
Tungsten halogen	40–50,000	4–2000	11–25	2700	100	1500–5000
Compact fluorescent (integral ballast)	100–5600	3–80	33–74	2700–6500	80–82	6000–15,000
Compact fluorescent (non-integral ballast)	250–9000	5–120	50–88	2700–6500	80–90	8000–20,000
Linear fluorescent	120–8850	6–120	20–105	2700–7500	50–98	8000–28,000
High-pressure sodium	3300–56,500	50–400	70–140	2000	25	28,500–60,000
White high-pressure sodium	1800–5000	45–115	40–50	2500	83	6000–9000
Ceramic metal halide	1700–41,000	20–400	83–110	3000–4400	78–93	6000–18,000

* LEDs are manufactured in a wide range of sizes and types; luminous efficacy and lifetime will vary with size and type. Typical examples of LEDs (as of September 2011) were chosen here.

† This is the lifetime to L_{70}, ie the mean time to diminish to 70% of the initial light output, at which point output is no longer acceptable.

‡ 'Lifetime increases to 32,000 hours at 400 cd/m² but light output not sufficient for general lighting applications'. Presentation to the Cambridge Enterprise and Technology Club (CETC), April 2008.

cd – candelas. K – kelvin. lm – lumens. nm – nanometre. N/A – not applicable. W – watts.

3   THE SUPPLY CHAIN

LED luminaire manufacturers currently supply a relatively small range of products for niche applications, often where lamp replacement is difficult or costly. It is likely to be some time before LEDs enter the mainstream wholesale market and the wide-scale replacement of general fluorescent lighting is implemented, as quality LED products are currently too expensive to bear the required levels of mark-up applied within the lighting industry. Consequently, cheap, poor-quality imports from the Far East have tended to give LED lighting a bad reputation in the wholesale market.

The solid-state lighting supply chain typically comprises the following:

1. **Semiconductor materials, substrate and encapsulation:** raw materials suppliers provide the semiconductor materials, substrates (such as sapphire or silicon carbide), encapsulants (ie clear materials to protect the chip and improve the light output) and phosphors (typically to convert blue light into white light). These materials have to be of high quality and may require special manufacturing processes.

2. **LED chips and devices:** an LED device manufacturer produces the LED chip from the raw materials, encapsulates it and adds a phosphor if required. Electrodes are mounted onto the material. This forms a basic LED device[5]. Alternatively, several LED chips may be packaged together to form an LED array. Production of the LED devices is a sophisticated manufacturing process requiring expensive semiconductor production plant. Nearly all of this manufacture is carried out outside the UK by large semiconductor manufacturers.

3. **LED drivers:** an electronics manufacturer produces the driver, an electronic device that converts the power supply into the required voltage and current for the LED. Drivers can be produced using standard electronics production plant by a large number of traditional semiconductor integrated chip manufacturers.

4. **LED light engines:** an LED lighting manufacturer produces the light engine (ie a combination of one or more LED devices or arrays), a driver, heat sink and electrical and mechanical connections. Light engines are intended to be mounted in a luminaire.

5. **Luminaires:** a luminaire manufacturer puts the LED light engine in a protective casing, which may also include optical components to change the distribution of light, and mechanical components to allow the luminaire to be mounted, positioned and aimed. Often the luminaire could be produced by a general lighting manufacturer buying in lamp and driver components, or a complete LED light engine.

6. **Wholesalers:** a wholesaler or importer sells the light fitting. They may keep the original manufacturer's name on it or rebadge it. There are a wide range of firms in this area ranging from the UK sales offices of foreign manufacturers through large wholesalers and down to very small firms of importers or suppliers. Currently most white light products are sold directly by manufacturers to end users or via specifiers.

7. **Specification and design:** a lighting designer specifies the light fitting and how it is to be used in the lighting scheme. Lighting designers may work independently or in small firms.

8. **Installation:** an electrician installs the light fitting.

9. **Maintenance and replacement:** staff replace the LED or complete luminaire if it is faulty or if its output has dropped below acceptable levels.

Some manufacturers combine various elements of the supply chain, eg combining stages 6 and 7 to operate a design and supply service. Alternatively some omit stage 7 altogether, especially in domestic or small-scale commercial buildings.

As LEDs are both an electronics product and a lighting product, the supply chain is currently divided into two parallel streams based around electronics and lighting. Historically electronics manufacturers have been dominant, but now more and more lighting companies are becoming involved, and the solid-state lighting market is becoming increasingly integrated into the mainstream lighting market.

Costs per unit drop substantially with larger-scale production, especially if manufacture is outsourced to the Far East or Eastern Europe. Often the early stages of manufacture will be UK based, but later stages may be outsourced abroad.

3.1   MARKET PROJECTIONS

Projections of the LED and OLED market suggest that by 2020 UK firms could have solid-state lighting sales of around £270 million, rising to £610 million by 2030 and £790 million by 2050. Profits are estimated to rise from £13 million in 2020 up to £36 million by 2030. (Profits are expected to plateau after 2030, despite the increasing market, because of competition from abroad and lower margins.)

OLEDs are expected to rise in efficacy from current values of 35 lm/W to 150 lm/W by 2014, then to stay at roughly this level. Unit costs of LEDs (in £ per light output) are expected to rival those of conventional lamps by 2015. OLEDs are expected to reach this level by 2020.

4 STANDARDS AND REGULATORY ISSUES

4.1 LIGHT QUALITY METRICS

LED integral ballast (self-ballasted) lamps tend to be quantified by the metric appropriate for the conventional lamp type that they are designed to replace. Therefore, for non-directional lamps designed to replace tungsten filament or fluorescent lamps, LED lamps are characterised by their total light output (luminous flux in all directions), measured in lumens, and the power consumed by the lamp as a whole, measured in watts. The energy efficiency of the lamp is sometimes also characterised by the luminous efficacy, which is simply the lumen output divided by the lamp power, given in lumens per watt. The higher the luminous efficacy, the more energy-efficient the lamp is. LED luminaires are often characterised in a similar way, but the lumen output is measured for the whole fitting and the power consumed is measured for the whole unit including the driver.

A few manufacturers quote 'chip efficiency', which can be misleading. Chip efficiency is the instantaneous lumen output of the LED chip divided by the power consumption measured without all the packing and/ or luminaire needed to make the LED a useful lighting product, and is not indicative of the performance that could be expected over sustained use. Such a figure is at least 80% higher than the useful luminous efficacy and is not a metric to predict the performance of the actual lamp or luminaire.

Directional LED lamps and luminaires designed to replace reflector tungsten halogen lamps will sometimes quantify their performance using the metrics that are commonly used for those lamps. The peak intensity is quoted, in candelas per m², together with a beam angle diagram showing the lux levels at various distances from the lamp. In some cases a polar diagram may also be given, which shows the proportion of the light emitted at various angles (Figure 6).

The colour (whiteness) of the LED is usually quoted in terms of correlated colour temperature (CCT). Products range from warm, yellowish whites with CCT of ~2700 K (which imitate tungsten filament bulbs and are popular for domestic use), through neutral whites in the 3000–3500 K range (which are popular for non-domestic use), to blueish whites with CCT of ~6500 K (which imitate daylight). As blueish white LEDs tend to be more efficient, many of the early products were of very high colour temperature (6000–7000 K). This colour is acceptable for exterior lighting but not very suitable for domestic applications. Using a warmer colour temperature will restrict the efficacy of the product as additional red phosphors are required, which are not as efficient at converting the blue light output of the LED chip as the yellow phosphors used in the more efficient lamps.

Another metric frequently quoted for lighting products is the colour rendering index (CRI). The CRI is intended to tell the consumer how accurately the light source will reproduce colours. Lamps with excellent colour rendering properties have a CRI between 90 and 100, and are often specified in situations where accurate colour judgement is required, such as in shops. The warmer white LEDs that have additional red phosphors to increase the warmth

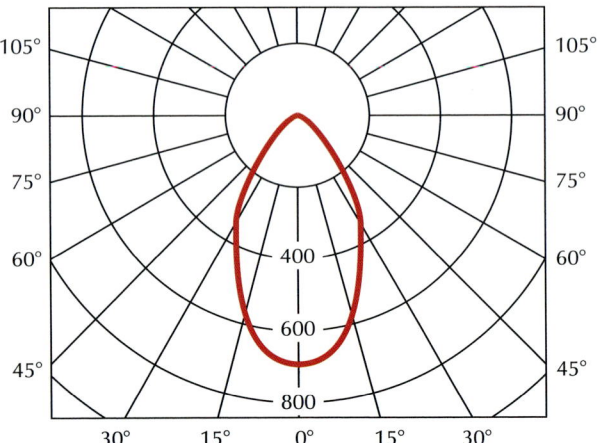

Height	Illuminance, lux (centre beam)	Beam diameter (m)
0.5	2800	0.58
1.0	700	1.15
1.5	311	1.73
2.0	175	2.31
2.5	112	2.89
3.0	79	3.46
Beam angle: 60°.		

Figure 6: Polar diagram and beam angle diagram for a typical directional lamp/luminaire

of the light usually have higher CRI values of more than 80. This puts them on par with light sources such as compact fluorescent lamps and tri-phosphor linear fluorescent lamps with regard to colour rendition. A few LED lamps will have colour rendering indices above 90; this excellent colour rendering is essential for some applications such as medical examinations and critical colour comparisons. However, the most efficient LEDs often have only fair CRI values (60–70) and will not show all colours well.

US researchers[6] have suggested that the CRI is not entirely suitable for LED spectra, and have proposed an alternative colour quality scale (CQS)[7]. An experiment was set up at BRE in 2009 to determine preferences for different LED lamp colours[8]. The experiment involved two sets of two booths containing everyday coloured objects, lit by different lamps (Figure 7). One pair of booths contained lamps equivalent in illuminance to 35 W halogen, and the other pair of booths contained lamps equivalent to 50 W halogen. Thirty-six subjects were invited to compare the brightness of each pair of booths, colour of the light and correctness or familiarity of the colours of the items.

One of the 50 W equivalents was a tuneable LED that had been specially set up to have a higher CQS rating (87) compared with its CRI (77). The colour of this lamp and the items lit by it were preferred to that of another LED lamp with a CQS rating of 83 and a CRI of 84. The difference in preference was statistically significant. The tuneable LED was even preferred to a tungsten halogen lamp with a CQS rating of 96 and a measured CRI of 99. The results suggested that for some LED lamps the CRI can be a poor indicator of colour preference.

4.2   LABELLING PROGRAMMES

LED luminaires are not currently subject to any specific labelling requirements. As the European Lamp Labelling Directive was introduced in 1998[9], before LED lighting was considered as a means of providing general lighting, it is not clear whether LED lamps are included in its scope. The wording of the scope is 'This Directive shall apply to household electric lamps supplied directly from the mains (filament and integral compact fluorescent lamps), and to household fluorescent lamps (including linear, and non-integral compact fluorescent lamps), even when marketed for non-household use'. This indicates the difficulty of writing these directives in a technology-neutral fashion while still being clear about which lamps are affected. What is clear is that lamps of input power less than 4 W are not included. The Lamp Labelling Directive is currently being reviewed by the European Commission and these problems of interpretation should be removed. In the meantime many manufacturers may choose to label their LED lamps to show that they have an energy efficiency rating of EU class A.

4.3   REGULATIONS

Non-directional lamps for household illumination are regulated by the European Commission Regulation 244/2009[10]. LED self-ballasted lamps designed to replace ordinary tungsten filament light bulbs are included within this legislation.

As the majority of these lamps will have an opal diffuser to disguise the individual LED modules and to distribute the light more evenly, they are categorised as

Figure 7: One pair of the experimental BRE test booths lit by lamps of different colour temperatures

'non-clear lamps' (Figure 8). In effect, this means that all such non-directional lamps sold in Europe must be of an efficacy equivalent to an energy efficiency rating of EU class A[9] where the lamp rated power must be lower than a maximum value related to luminous flux.

This regulation does mean that all LED lamps being sold as non-directional household bulbs should be at least as efficient as a compact fluorescent lamp, which should prevent a proliferation of poor-quality lamps on the European market. However, it also acts as something of a barrier as lamps must use very good quality LED chips in order to achieve this performance level, which means that they are very expensive.

4.4   TESTING

There are no measurement standards specific to LED luminaires at present. Instead, the measurement standard for lamps and luminaires, BS EN 13032-1:2004[11], is used as a de facto standard for the testing of LED luminaires by most UK test houses. This standard does include some detail about measurement of luminous flux but also refers back to CIE 84[12], a publication that is more than 20 years old.

In North America, the Illuminating Engineering Society (IES) has published an 'approved method' for measuring the electrical and photometric properties, IES LM-79-08[13]. This approved method allows the use of either an integrating sphere or a goniophotometer for the measurement of luminous flux.

Both these methods have common ground and both documents are approved by the Enhanced Capital Allowances (ECA) scheme[14] as suitable test procedures.

Measuring lumen maintenance after 6000 hours is a pragmatic approach to providing an indication of performance over what can be very long lifetimes for LED products. Six-thousand hours still represents over nine months of testing time, but very long-lived LED products will scarcely show any measurable reduction in luminous flux over that period. As LED light sources tend to diminish in light output with time rather than to fail catastrophically like conventional light sources, the lumen maintenance is often quoted in terms of an L_{70} value, ie the mean time to diminish to 70% of the initial

light output. When the light output falls below 70% of its initial value the lamp is unlikely to be providing sufficient illumination for the required task. Many LED products are quoted as having L_{70} lifetimes of 35,000–50,000 hours (or more), which equates to losses of 5% and 3.6% over 6000 hours, respectively (assuming linear degradation). As the measurement uncertainty is in the order of 5%, these losses are barely discernible even after 6000 hours.

IES has also published an approved method for the determination of LED lumen maintenance (lifetime), IES LM-80-08[15]. This publication also suggests testing for a minimum of 6000 hours with 1000-hour intervals, although it recommends 10,000 hours of testing for more reliable prediction of future performance. It suggests that rated lumen maintenance life can be quoted either to 70% (L_{70}) or 50% (L_{50}) lumen maintenance. However, a 50% reduction in light output would be very noticeable and almost certainly would be unacceptable for general lighting purposes.

4.5   PERFORMANCE SPECIFICATION

There are currently two specifications for LED performance being used in the UK: the Energy Saving Trust Recommended (ESTR) criteria[16, 17] and the Enhanced Capital Allowances (ECA) criteria[14].

The ESTR scheme has been developed for domestic applications. The criteria apply to LED lamps and external luminaires. The latest version of the criteria (version 2.0) for LED lamps gives eligibility criteria for all types of LED lamp and provides best-practice levels for lamps intended for the domestic lighting market, such as those used in wall fittings (Figure 9).

The ECA scheme enables buyers of energy-efficient equipment to claim the cost against tax in the year of purchase. Criteria have been developed by the Carbon Trust to represent a best-practice standard for commercial applications. The scheme currently covers white LED lighting units for amenity, accent and display purposes, as well as white LED units for general interior lighting and for exterior area lighting and exterior floodlighting (as of September 2011).

Figure 8: Miniature LED lamps with different diffusers

Figure 9: An LED wall light fitting

Individual standards for LED lamps are listed in a Lighting Industry Federation Technical Statement[18]. These standards either specialise in one particular aspect of LED performance or safety, or they are general standards that apply to all lamps, not just LEDs.

If the LED lamp or fitting is CE marked, it should comply with all the relevant safety standards. Some aspects of LED luminaire performance will be covered by the general luminaire performance standards, ie the relevant part of BS EN 60598[19]. Safety specifications for LED modules are covered by BS EN 62031:2008[20].

A performance standard is being developed for self-ballasted LEDs, IEC/PAS 62612:2009[21]. This has reached the stage of being a Publicly Available Standard but is incomplete and still subject to change before it is fully adopted. For photometric measurements it refers to CIE 84[12], although it does warn that CIE 84 is not optimised for LED lamps. It defines tolerances on various measurements including colour temperature and CRI and suggests a category system for rating lumen maintenance after a maximum of 6000 hours.

In IEC/PAS 62612:2009[21], Category A lumen maintenance lamps (ie the best category) must have a measured flux after 6000 hours that has decreased by not more than 10% of rated flux. This equates to a minimum L_{70} value of 18,000 hours.

Drivers for LEDs are covered by two standards: BS EN 62384:2006[22], which covers performance requirements, and BS EN 61347-2-13:2006[23]. In some cases the driver would also be covered by the standard for plug-in transformers, BS EN 61558-1:2005[24].

5   ADVANTAGES AND BENEFITS OF LEDS

LEDs offer enormous possibilities for providing highly energy-efficient and good-quality lighting. Their properties can vary greatly from one manufacturer to another in terms of light output and colour quality. However, careful specification will ensure that LED lighting meets the requirements of the particular application.

5.1   COSTS

LED costs have dropped dramatically since their conception and are expected to continue to fall as improvements in technologies such as automation and large-scale manufacturing increase LED output and efficacy. The capital cost of LEDs is currently around seven times that of compact fluorescent lamps, but lifetime costs are already comparable to those of incandescent and halogen lamps owing to the much lower power consumption and longer life of LEDs.

Considerable effort is being made to improve the overall luminous efficacy of LED chips to achieve efficacy improvements in line with the projections published yearly by the US Department of Energy (DOE)[25], so that lamps and luminaires using the latest technology will always be sold at a premium price and products using 'older technology' will quickly reduce in price.

Outputs from a US DOE workshop held in April 2010 set out the anticipated reductions in manufacturing costs of LED luminaires to 2020[5]. The 2009 cost was taken as a benchmark figure and a consensus reached as to how this cost was split across the manufacturing process. The outcomes highlight an anticipated reduction in manufacturing costs of 78% over the 2009 benchmark by 2020, with the LED packages no longer making up the largest portion of the costs; to be taken over by the mechanical/thermal element of the luminaire. These values are extrapolated in Figure 10 below.

The cost of the mechanical/thermal part of the luminaire is the most difficult to reduce. It includes both the physical components – comprising the complete luminaire fixture and means for mounting the LED(s), driver and optical components – and the thermal components, which are required to remove the heat produced within the fixture. As the thermal components are always likely to include a substantial amount of metal[26], and as the cost of metals is unlikely to decline in the future, this could be one limiting factor in reducing cost. The US DOE suggests that steps need to be taken to actively address cost-reduction strategies for this part of the luminaire[5]. New materials such as phase-change materials may provide an alternative to large volumes of metal for thermal management.

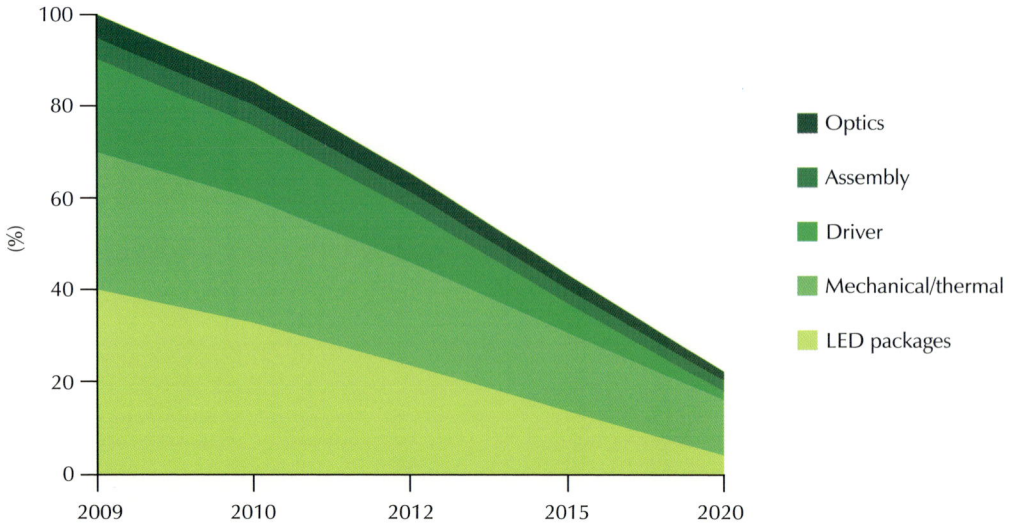

Figure 10: Anticipated relative reductions in manufacturing costs of LED luminaires to 2020 (2009 benchmark = 100)

An analysis was also carried out on the relative manufacturing costs of the LED package alone. Again 2009 was taken as the benchmark and a consensus reached as to how the costs are split according to the individual parts of the LED package. The outcomes highlight an anticipated reduction in the cost of the LED package by 89% over the 2009 benchmark by 2020. Throughout this period packaging remains the largest portion of the costs, which all reduce in approximately the same ratio (Figure 11).

According to the US DOE, 'Though not reflected in the cost projection, improvements in an earlier part of the manufacturing process, such as improved uniformity in the epitaxial process, will have a "lever" effect and can greatly impact the final device cost and selling price through improved binning yields'[5].

At present, the best-quality LED products use sapphire substrate discs of relatively small sizes (typically 50–100 mm). They are expensive to produce due to their sensitivity to deformation, which can occur more readily than on other substrates, and therefore represent the main cost of LED packages. Bulk cheap solid-state devices are usually produced on larger (150 mm or 200 mm) silicon or silicon carbide wafers; producing good-quality LEDs on this substrate is a technical challenge. Larger sapphire (150 mm) discs[27] and silicon discs (150–200 mm)[28] have been produced very recently and will help to bring down the cost of the LED packages.

5.2   SUPPORT AND FUNDING

There are currently two main types of subsidy or incentive applicable to lighting: one for domestic lighting products and one for non-domestic products.

5.2.1   Subsidies for domestic lighting products

The most efficient LED lamps and luminaires, ie those conforming to ESTR[16] criteria, are eligible to be considered by public energy supply companies for subsidising under the Carbon Emissions Reduction Target (CERT)[29] scheme (2008–2011). This scheme has been very successful in reducing the market cost of compact fluorescent lights but it is unclear how much (if any) impact it will have on helping to bring down LED lamp prices. The difficulty is that LED lamps can no longer be compared with tungsten filament lamps as these are being removed from the market. Very few LED products have yet received ESTR accreditation to allow them to be eligible for consideration by CERT.

5.2.2   Incentives for non-domestic lighting products

The most efficient LED luminaires are eligible under the ECA scheme to receive a tax incentive. In 2010 eligibility was extended from 'amenity, accent and display' lighting to include general lighting applications. The luminous efficacy requirements for eligible luminaires are high: 46 lm/W for amenity, accent and display lighting and 60 lm/W for general lighting. ECAs allow companies to offset the full cost (plus installation and transport) of energy-efficient products against their corporation or income tax liabilities in the first year rather than having to write it off over 10 years. This provides a cash flow boost of more than 20p for every £1 invested.

5.2.3   Research grants

The UK government supplies research grants through the Engineering and Physical Sciences Research Council (EPSRC) and the Technology Strategy Board (TSB); these grants are potentially available for future development of LED lighting technologies. Building regulations and future restrictions on available lamp types via the Energy-Using Products Directive[30] may also enhance the uptake of LEDs in future.

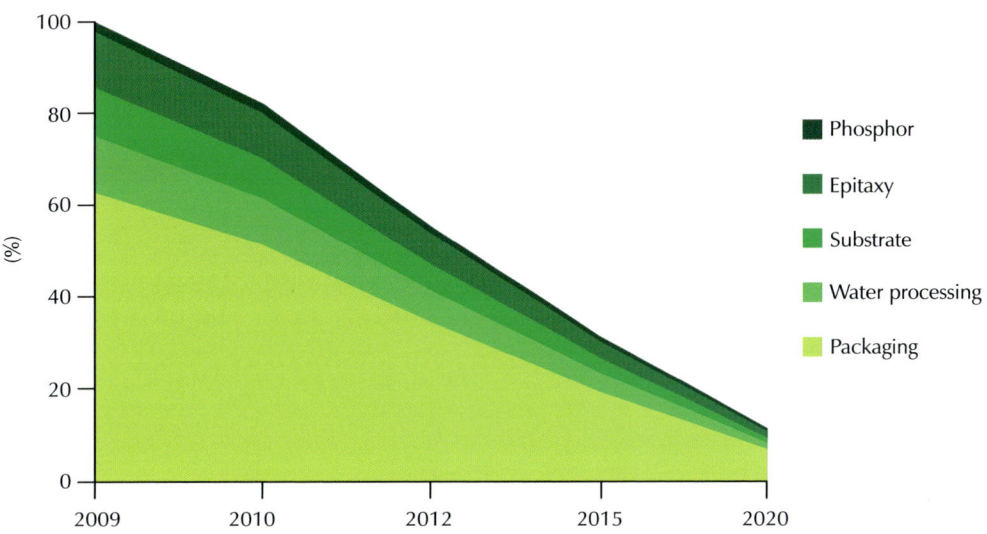

Figure 11: Anticipated relative reductions in manufacturing costs of LED packages to 2020 (2009 benchmark = 100)

5.3   RESEARCH AND DEVELOPMENT

The UK is well endowed with expertise throughout the solid-state lighting supply chain, including leading research groups at a number of universities. Current research and development efforts are mainly focusing on the material properties of the basic LED devices themselves and also on reducing the cost of LEDs through new manufacturing techniques, improving phosphors for white LEDs, getting all the available light out of the LED module and encapsulating OLEDs to protect them from moisture. However, these devices are nearly all made abroad; UK manufacturing expertise is concentrated further down the supply chain, in producing LED light engines and luminaires.

The University of Sheffield carries out state-of-the-art semiconductor epitaxy and device fabrication using III-V semiconductors. It also carries out research into advanced technologies for device fabrication, and custom optoelectronic and electronic devices. Most of this research is with academic groups, supported by grants from the EPSRC, Biotechnology and Biological Science Research Council (BBSRC), Ministry of Defence (MOD), TSB, and EU and industrial collaborations.

The Wide Band Gap Semiconductors Group at the University of Bath is interested in many aspects of the physics and technology of wide band gap materials and especially the group III nitrides, and recently received a TSB grant for a project, 'Novel LEDs for efficient lighting systems'.

Research at the University of Cambridge has focused on gallium nitride as an LED material. The Cambridge Centre for Gallium Nitride has strong academic collaborations with Manchester University, the University of Oxford and Sheffield Hallam University, and strong industry collaborations with Thomas Swan, AIXTRON, Forge Europa and QinetiQ.

The University of Nottingham's Nitrides Research Group was formed in 1991, and established what is still the only plasma-assisted molecular beam epitaxy growth facility in UK academia for the complete range of group III nitrides.

The University of Strathclyde's Institute of Photonics has been active in gallium nitride growth and optoelectronic device fabrication for the past 10 years.

The Phosphor Group at Brunel University was established in 1970 and is the only academic group in Europe or the US that has carried out research on phosphors over this period.

In terms of future research and development, a US DOE workshop identified a number of manufacturing research and development priority tasks (Table 4). In addition to the areas identified in Table 4, under the EU Waste Electrical and Electronic Equipment Directive[31] there is a responsibility to recycle LEDs. LEDs contain small quantities of rare elements like gold, silver, indium and gallium. Most of them also contain larger quantities of aluminium in heat sinks, and copper wiring. However, LEDs are not generally easy to disassemble and large-scale recycling facilities have yet to be developed. This is likely to become a more important issue in the future as the LED market grows and LEDs start to reach their end of life.

5.4   FUNCTIONALITY

Most LEDs can operate using conventional analogue (1–10 V) or digital (eg DALI, ESB, DSI) control systems. Depending on the type of LED, they can be entirely free from flicker, automatically switched, dimmed and in some cases the colour can be changed. This last possibility has already been used extensively in decorative exterior and theatre lighting and in mood lighting in clubs, pubs and retail outlets, as well as in some homes.

Controllable LEDs open up the possibility of smart environments, where monitoring and home automation help extend the time that the elderly can stay in their home while maintaining their normal lifestyle[32]. A large component of the smart home environment is lighting, and the potential of LEDs far exceeds that of other technologies to meet the needs of the ageing population.

Circadian lighting (to match patterns within the body's daily cycle) is an obvious place where LEDs – by being able to give a unique spectral prescription of light – can help normalise the sleep patterns of older adults, increase their sense of well-being and be able to reduce the symptoms of Alzheimer's disease. Research at the Lighting Research Centre of Rensselaer Polytechnic University[33]

Table 4: LED manufacturing R&D priority tasks[5]

Task	Description
Luminaire/module manufacturing	Automation, manufacturing or design tools to demonstrate high-quality, flexible manufacturing at low cost
Driver manufacturing	Improved design for manufacture for flexibility, reduced parts count and cost, while maintaining performance
Test and inspection equipment	High-speed, non-destructive and standardised equipment for all manufacturing steps
Tools for epitaxial growth	Tools, processes and precursors to lower cost of ownership and improve uniformity
Wafer-processing equipment	Tailored tools to improve LED wafer processing
LED packages	Improved back-end processes and tools to optimise quality and consistency and to lower cost
Phosphor manufacturing and application	High-volume phosphor manufacture and efficient materials application

suggests that not just elderly people but all segments of society – especially in the era of deep-core offices – can benefit from appropriately-timed doses of light.

LEDs require a fixed voltage in order to operate correctly; therefore dimming can only be achieved by cutting the mains supply. High-quality thyristor dimmers use an electronic circuit to cut the mains supply. This means that the power is applied 100 times per second, for a period between 0 and 1/100th of a second, as set by the control[34]. This kind of dimmer is compatible with 'dimmable' lighting transformers for low-voltage alternating current (AC) systems. Because full power is applied to the lamp for part of the time, these dimmers will work with some kinds of LED product. Those where the power is applied directly to the chain of LEDs, such as 230 V ropelight, will also usually respond to this style of dimming. When the dimming gets below 50% the remaining AC cycle being applied to the lamp will not be sufficient to power it and it will quickly turn off. LED products where the power is further conditioned will not be affected by these dimmers; this is usually the case with 230 V and 12 V bulbs.

Another dimming method frequently used is pulse width modulation. The full direct current (DC) voltage is applied to the product for a very short time, thousands of times per second. The LEDs effectively flash, too quickly for the eye to see. The resultant brightness depends on the time the power is applied. This kind of dimmer should successfully dim – all the way from full to off – most LED products designed to operate directly from DC voltages. Again, if there is further power conditioning included within the product then this is likely to negate the dimming effect. Pulse width modulation is inherently incompatible with AC products, although thyristor dimmers offer a similar solution for AC products.

Most high-power LEDs are designed to operate from constant current sources. The supply will apply the voltage necessary to achieve the required current. If this current can be varied the supply can be used as a dimmer.

5.5   DRIVER REQUIREMENTS

Drivers can be produced using standard electronics production plant although a key issue is the size of the driver: very small drivers are more attractive where luminaire size is limited. Recent advances in driver-integrated circuits will allow many companies to design their own intelligent drivers with optimised feature sets such as dimming, colour control and sensors.

A US DOE manufacturing workshop[5] identified 'the need for drivers with improved design for manufacturing, integration, and flexibility within the luminaire … [which] could include the disaggregation of driver functionality into sub-modules to allow luminaire integrators to mix and match functions while maintaining high efficiency and reliability'.

The size and 'form factor' (ie shape) of drivers is also important, especially for integration into lamps or small luminaires. Driver performance can be affected by the load placed on them and by thermal conditions. The quality of components is vital to avoid the driver being the limiting factor in LED luminaire lifetime. For instance, electrolytic capacitors can easily fail significantly before the LED chip has reached the end of its useful life.

LEDs can change in colour over their lifetime, particularly if they are made up by mixing red, green and blue sources. Some LED light engines have special drivers that sense departures in colour and rebalance the light output of the different coloured sources. Such tuneable LEDs can also be used for special effects, creating a wide range of different colours for display and entertainment applications. Figure 12 shows LED lighting being used for medical purposes due to its good colour rendering properties.

The US DOE workshop also identified a need for standardised information on driver performance to facilitate driver integration into LED-based luminaires. Proposed driver information comprised:
- Compatibility with ambient light sensors
- Compatibility with specific dimming protocols
- Efficiency with respect to power, load and temperature
- Harmonic distortion in power supply
- Input voltage and output voltage variation
- Maximum output power
- Off-state power
- Operating temperature range
- Output current variation with temperature, voltage, etc.
- Power factor correction
- Power overshoot
- Power-to-light time
- Transient and overvoltage protection specifications

Figure 12: LED lighting with good colour rendering for medical purposes

6 CHALLENGES AND BARRIERS TO ADOPTION

Evidence suggests that the UK market may be receptive to new light sources provided these sources meet customer requirements in terms of cost, size, colour and perceived brightness. Longer-term barriers could include a lack of fabrication capacity worldwide and a shortage of raw materials such as gallium. There are also issues surrounding the credibility of new technologies, compatibility with existing fittings and reluctance of traditional manufacturers to accelerate uptake.

Traditionally manufacturers have tended to produce a standard range of types of luminaire, sometimes using interchangeable lamps. For example, the same luminaire might be available in tungsten or metal halide or compact fluorescent types; only the lamp and control gear are different. This type of approach tends not to work well with many applications of LED because of the intrinsically directional properties of the light source and the need for thermal control, conducting heat away from the device.

Uniformity of LED chip production has been challenging and the colour temperature of chips has varied over a single substrate disc. This has led to the need to sort the chips into colour bins, which is a costly and time-consuming process. Specifying chips from a narrow bin width will give greater uniformity of colour, but this advantage needs to be offset against the higher cost of a tight bin specification. Uniformity has been improved somewhat by a variety of new techniques, including (i) the use of a reflective bucket containing an encapsulant into which the chip is mounted, which keeps the phosphor layer remote from the chip; and (ii) the use of a controlled phosphor disc that is matched and mounted in close contact to a thinned sapphire substrate, which becomes the top of the chip with the electrical contacts attached below.

In addition to the challenges surrounding their manufacture, each LED will also ultimately require disposal. Large-scale recycling facilities have yet to be developed to support their effective disposal and enable the recovery of copper and aluminium together with the smaller quantities of rare elements (eg gold, silver, indium, gallium) to be recovered. This is likely to become a more important issue in the future as the LED market grows and LEDs start to reach their end of life.

At the current time, OLEDs have a similar problem to LEDs in producing stable white light and high efficacy and are also very sensitive to moisture. Unlike LEDs, the green emission of OLEDs can be considerably more efficient than for either red or blue emission. Although efficacy of OLEDs is good, they currently have too short a lifetime to be considered suitable for general lighting applications. They are currently used commercially in the displays of mobile phones and similar devices.

7   CONCLUSIONS AND RECOMMENDATIONS

Whereas lighting is responsible for between 15% and 22% of all electricity use in buildings, LEDs offer enormous potential as compact, low-energy sources for providing highly energy-efficient and good-quality lighting.

Since the production of the first white LED in 1996, considerable effort has been made to improve the efficacy of LED and OLED lighting and to reduce the costs of manufacture. If LED lighting achieves its expected levels of efficiency, then with high levels of uptake the energy consumption of domestic and commercial lighting could potentially be reduced by up to 70% by 2050. It could realistically achieve a 37% saving in lighting energy use by 2030.

LEDs are currently mainly used in niche applications, such as: coloured decorative lighting; addressable picture walls; emergency lighting, automotive and aviation lighting; low-power display lighting; and LCD backlighting. Recent applications of LEDs include vehicle headlights, traffic signals, downlights (primarily for commercial accent and display applications), emergency lighting and decorative outdoor lighting, particularly in conjunction with small-scale solar panels.

Projections of the LED and OLED market suggest that by 2020 UK firms could have solid-state lighting sales of around £270 million, rising to £610 million by 2030 and £790 million by 2050. Profits are estimated to rise from £13 million in 2020 up to £36 million by 2030.

Compared with other forms of lighting, LEDs tend to have lower light output and lower wattage: typically a few watts with currents in the order of milliamps or tens of milliamps. The best-performing commercially available warm white LED fittings have an efficacy of 50–60 lm/W. Within a few years, it is expected that the efficacies of LED lamps will rise to 100 lm/W or above (the highest-efficiency high-power white LED already achieves 115 lm/W), so lower-wattage lamps may then be able to provide the required amounts of light. OLEDs are expected to rise in efficacy from current values of 35 lm/W to 150 lm/W by 2014.

Although LED luminaires are not currently subject to any specific labelling requirements, many manufacturers may choose to label their LED lamps to show that they reach an energy efficiency rating of EU class A. However, European Commission Regulation 244/2009 regarding non-directional lamps for household illumination covers LED self-ballasted lamps designed to replace ordinary tungsten filament light bulbs and requires that all such LED lamps sold in Europe must reach the EU class A energy efficiency rating.

Besides luminous efficacy, the other LED properties vary considerably, so careful specification is required. This needs to cover the amount and direction of the output light, glare, lifetime and colour quality (including colour appearance, colour rendering and colour constancy between batches and over time). LEDs tend to perform poorly at high temperatures, so their fittings require heat sinks or ventilation to keep the LEDs cool.

8 REFERENCES

1   Navigant Consulting Inc. Energy savings potential of solid-state lighting in general illumination applications. Washington DC, US Department of Energy, 2006.

2   Gather M, Köhnen A, Meerholz K, Becker H and Falcou A. Solution-processed full-color polymer-OLED displays fabricated by direct photolithography. SID Digest 2006 (P-181), 37 (1) 909–911.

3   Navigant Consulting Inc., Radcliffe Advisors Inc. and SSLS Inc. Solid-state lighting research and development. Multi-Year Program Plan FY'09–FY'14. Washington DC, US Department of Energy, 2008.

4   LEDs Magazine. Osram reports warm-white OLED with 46 lm/W efficiency. March 2008.

5   Bardsley Consulting, Navigant Consulting Inc., Radcliffe Advisors Inc., SB Consulting and SSLS Inc. Solid-state lighting research and development: manufacturing roadmap. Washington DC, US Department of Energy, 2010.

6   Davis W and Ohno Y. Toward an improved color rendering metric. Proceedings of SPIE: Fifth International Conference on Solid State Lighting. 2005 (5941) 59411G.

7   Commission Internationale de l'Eclairage (CIE). Colour rendering of white LED light sources. CIE publication 177:2007. Vienna, CIE, 2007.

8   Littlefair P and Graves H. Specifying LED lighting. BRE IP 15/10. Bracknell, IHS BRE Press, 2010.

9   Commission Directive 98/11/EC. Commission Directive 98/11/EC of 27 January 1998 implementing Council Directive 92/75/EEC with regard to energy labelling of household lamps. Brussels, European Parliament and Council of the European Union, 27 January 1998.

10 Commission Regulation 244/2009. Commission Regulation 244/2009 of 18 March 2009 implementing Directive 2005/32/EC of the European Parliament and of the Council with regard to ecodesign requirements for non-directional household lamps. Brussels, European Parliament and Council of the European Union, 18 March 2009.

11 BSI. Light and lighting – Measurement and presentation of photometric data of lamps and luminaires. BS EN 13032-1:2004. London, BSI, 2004.

12 Commission Internationale de l'Eclairage (CIE). The measurement of luminous flux. CIE publication 84:1989. Vienna, CIE, 1989.

13 Illuminating Engineering Society (IES). Electrical and photometric measurements of solid-state lighting products. LM-79-08. New York NY, IES, 2008.

14 Energy Technology List. About ECA. Available at: www.eca.gov.uk/etl/about/_about.htm.

15 Illuminating Engineering Society (IES). Measuring lumen maintenance of LED light sources. LM-80-08. New York NY, IES, 2008.

16 Energy Saving Trust (EST). LED requirements for replacement lamps and modules. Version 2.0. London, EST, 2009.

17 Energy Saving Trust (EST). LED luminaire requirements. Version 2.0. London, EST, 2009.

18 Lighting Industry Federation (LIF). Key standards for production, testing and measurement of LED-based luminaires. LIF Technical Statement 44. London, LIF, 2009.

19 BSI. Luminaires. BS EN 60598. London, BSI.

20 BSI. LED modules for general lighting – Safety specifications. BS EN 62031:2008. London, BSI, 2008.

21 BSI. Self-ballasted LED lamps for general lighting services – Performance requirements. DD IEC/PAS 62612:2009. London, BSI, 2009.

22 BSI. D.C. or A.C. supplied electronic control gear for LED modules – Performance requirements. BS EN 62384:2006. London, BSI, 2006.

23 BSI. Lamp controlgear – Particular requirements for D.C. or A.C. supplied electronic controlgear for LED modules. BS EN 61347-2-13:2006. London, BSI, 2006.

24 BSI. Safety of power transformers, power supplies, reactors and similar products – General requirements and tests. BS EN 61558-1:2005. London, BSI, 2005.

25 US Department of Energy. Solid-state lighting: technology roadmaps. Available at: www1.eere.energy.gov/buildings/ssl/techroadmaps.html.

26 Navigant Consulting Inc. Life cycle assessment for ultra-efficient lighting. London, Department for Environment, Food and Rural Affairs (Defra), 2009.

27 LEDs Magazine. Rubicon quarterly financials portend a transition to 6-in LED wafers. August 2010.

28 LEDs Magazine. Azzurro transfers GaN-on-Si technology to Osram. November 2009.

29   Department for Energy and Climate Change (DECC).
     Consultation on proposed amendments to the Carbon Emissions
     Reduction Target 2008–2011. Available at: www.decc.gov.uk/
     en/content/cms/consultations/open/cert/cert.aspx.

30   Directive 2005/32/EC. Directive establishing a framework
     for the setting of ecodesign requirements for energy-using
     products and amending Council Directive 92/42/EEC and
     Directives 96/57/EC and 2000/55/EC of the European
     Parliament and of the Council. Brussels, European Parliament
     and Council of the European Union, 6 July 2005.

31   Directive 2002/96/EC. Directive on waste electrical and
     electronic equipment (WEEE). Brussels, European Parliament
     and Council of the European Union, 27 January 2003.

32   LEDs Magazine. LED lighting can enable smart homes and
     enhance lifestyles. April 2010.

33   Figueiro M G, Bullough J D and Rea M S. Spectral sensitivity
     of the circadian system. In: Proceedings of the Society of
     Photo-Optical Instrumentation Engineers. Bellingham WA,
     Society of Photo-Optical Instrumentation Engineers, Vol 5187,
     pp 207–214.

34   Exled. Dimming LEDs. Available at: www.led-lightbulbs.co.uk/
     main.asp?sitepages=Tech-Dimming-LEDs.

GLOSSARY OF TERMS AND ABBREVIATIONS

addressable OLED	an organic LED that allows individual pixels of material to be switched separately
candela	the SI unit for luminous intensity, which is the power emitted by a light source in a particular direction
colour quality scale (CQS)	a quantitative measure of the ability of a light source to reproduce colours of illuminate objects
colour rendering index (CRI)	a quantitative measure of the ability of a light source to reproduce colours of various objects in comparison with an ideal or natural light source
compact fluorescent	a type of fluorescent lamp, designed to replace incandescent lamps, which uses less power and has a longer rated life
controllable LED	an LED connected to a dimmable driver, which is able to vary the output current; LED colour changes are made possible by dimming separately three different channels of red-, green- and blue-coloured LEDs
correlated colour temperature (CCT)	a quantitative measure that defines a colour as the temperature in degrees Kelvin (K) that a 'black body' source must reach in order to produce that same colour
directional lamp	a lamp that contains reflectors that direct and control the light
efficacy/luminous efficacy	the amount of light (luminous flux) produced by a lamp, usually measured in lumens (lm), as a ratio of the amount of power consumed, usually measured in watts (W), to produce it; the ratio is usually expressed in lm/W
epitaxy	the method for depositing a monocrystalline film onto a monocrystalline substrate
goniophotometry	the technique used for measuring the angular distribution of light scattered from a surface
high-power LED	an LED that can be driven at currents from hundreds of milliamps to more than an ampere, compared with the tens of milliamps for other LEDs
incandescent lamp	a lamp that makes light by heating a metal filament wire to a high temperature until it glows
induction lamp	an electrodeless lamp, where an induced electromagnetic field, generated by induction coils, excites the mercury atoms in the glass tube, causing them to emit UV radiation that is converted to visible light by the phosphor coating on the inside of the tube
LED	a light-emitting diode; a semiconductor light source
LED array	an arrangement of multiple LEDs to form a lamp with higher light output and/or with colour-changing possibilities
LED chip	a slice of semiconducting material doped with impurities to create the p–n junction, where energy in the form of photons is emitted by electrons falling into lower energy levels after meeting holes
LED device	an LED that converts electrical energy into light
LED light engine	the light-producing portion of a light fixture, containing an array of individual LEDs mounted in configuration to disperse light in a designed pattern
LED package	the plastic body surrounding the LED
low-pressure sodium lamp	a type of lamp, commonly used in street lights, that produces a bright yellow light by causing the sodium metal within the tube to vaporise

lumen maintenance	the amount of light emitted from a source at any given time relative to the light output when the source was first measured; this is usually expressed as a percentage
lumen	the unit used to quantify the amount of light produced by a lamp (luminous flux)
lumens per watt (lm/W)	the unit used to quantify the efficacy of a lamp
luminaire	a light fixture or fitting; an electrical device used to create artificial light and/or illumination, by use of an electric lamp
luminous flux	the amount of light produced by a lamp, usually measured in lumens (lm)
metal halide	a type of high-intensity discharge lamp that produces light by means of an electric arc between tungsten electrodes
narrow bin width	a reduced, sharper extent in which parameters of different LED lamps are specified, in order to minimise differences in the same parameter between different lamps and to increase uniformity
non-directional lamp	a lamp not having at least 80% light output within a solid angle of π sr (corresponding to a cone with an angle of 120°)
OLED	an organic LED in which the emissive electroluminescent layer is a film of organic compounds that emit light in response to electric current
opal diffuser	a device used for diffusing or spreading light to create soft light
optoelectronic device fabrication	the process of manufacturing electronic devices that generate, detect and use in their operation visible light and invisible forms of radiation such as gamma rays, X-rays, ultraviolet and infrared
phase-change material	a substance with a high heat of fusion, which, melting and solidifying at a certain temperature, is capable of storing and releasing large amounts of heat
photometry	the science of measuring visible light in terms of its perceived brightness to human vision
point source	a single identifiable localised source of light having negligible size relative to other length scales
P-OLED	an OLED in which electroluminescent conductive polymers emit light when connected to an external voltage
pulse width modulator	an electronic device using the pulse width modulation technique to dim LEDs
self-ballasted lamp	a lamp of the arc-discharge type, which incorporates a current-limiting device
SM-OLED	a small-molecule OLED, where the emissive electroluminescent layer is made of materials with low molecular weight such as organometallic chelates, fluorescent and phosphorescent dyes and conjugated dendrimers
solid-state lighting	a type of lighting that uses semiconductor LEDs, OLEDs or P-OLEDs as sources of illumination rather than electrical filaments, plasma or gas
sulfur lamp	a highly efficient full-spectrum light that is generated by sulfur plasma that has been excited by microwave radiation
thyristor dimmer	a type of dimmer that uses switching techniques, which results in almost instantaneous dimming
tubular fluorescent lamp	a fluorescent lamp in a tubular form; light is produced when electricity excites mercury vapour, which in turn produces short-wave ultraviolet light that causes a phosphor to fluoresce, producing visible light
tuneable LED	an LED lamp capable of varying the colour of the emitted light
tungsten filament lamp	an incandescent lamp with a tungsten filament that emits light when the filament is heated
tungsten halogen lamp	an incandescent lamp with a tungsten filament contained within an inert gas and a small amount of halogen; the tungsten and halogen cause a chemical reaction that increases the lifetime of the lamp and prevents darkening seen in tungsten filament lamps
wide band gap material	a type of semiconductor that has electronic band gaps (energy range in a solid in which no electron states can exist) larger than one or two electronvolts (eV)

Other reports from BRE Trust